The Beekeepers Annual 2016

THE BEEKEEPERS ANNUAL
IS PUBLISHED BY
NORTHERN BEE BOOKS
MYTHOLMROYD,
WEST YORKSHIRE

PRINTED BY
LIGHTNING SOURCE, UK
ISBN 978-1-908904-82-9

MMXIV

EDITOR, JOHN PHIPPS
NEOCHORI, 24024 AGIOS NIKOLAOS,
MESSINIAS, GREECE
EMAIL manifest@runbox.com

SET IN HELVETICA LT BY D&P Design and Print

Cover: 'Sun Hive' (*Photo: Heidi Herrmann*)

The Beekeepers Annual 2016

CONTENTS

FOREWORD

John Phipps

September 2015

The theme of the main content of this edition of the Annual is about caring.

Many hobbyists are caring for bees in an alternative way to help them during these troubled times - the emphasis being to ensure their survival for the valuable pollination work they do; the chance of obtaining any honey harvest is of a very low priority. With the universal dwindling of honeybee populations, the application of this relatively new philosophy may well help to restore colony numbers, especially if bees are kept in a more natural way with little disturbance from the beekeeper. Having bees in a garden adds a new dimension to it, and the opportunity just to watch the bees at work amongst the flowers, or their activities at the entrance to the hives, will not only allow the onlooker to derive a great deal of pleasure from their presence, but those who are more observant and thoughtful will gain a good understanding of their charges. Sideliners and commercial beekeepers, whose main purpose is to produce honey, may well learn from this bee-centred focus on the craft, for careful husbandry and a more sensitive approach when handling colonies should produce greater rewards and satisfaction.

In the plates at the beginning of each month, examples of a dozen hives are illustrated, all of which allow beekeepers to keep their bees in a natural and sustainable way. Whilst not so long ago those beekeepers wishing to keep their bees in many of the hives described had to make them themselves, now, with a greater interest being shown, several of the models are being produced by the main beekeeping firms. I have not included in the examples any of the hive types which are constructed in plastic, as I think it is inappropriate to use anything but natural materials for the homes of bees.

Included, too, in this Annual, are two articles about beekeepers who have a great concern for those people who have social or personal problems or who are serving lengthy sentences within a prison. The introduction of bees into these peoples' lives is seen as having a profound effect on their behaviour. Through beekeeping many skills have been learned by them so their lives have become more enriched and meaningful. We all know that time spent with bees is therapeutic, so even if those with problems can escape from them for only a short time, then the efforts of those providing care and instruction are invaluable.

John Phipps
September 2014

Why are we having so many problems with our Queens?

PROBLEMS WITH QUEEN BEES:

many years of problems and still no solution . . .

Roger Patterson

I started beekeeping in 1963 and since 2002 I have been trying to highlight problems I have experienced with queens that I rarely saw until recently. I am a very experienced beekeeper who had 130 colonies at one stage. I am heavily involved with my local BKA and travel widely lecturing and demonstrating, so I speak to a lot of beekeepers and see a lot of colonies. I am not a new beekeeper with limited experience who has seen something a few times, or am confused by what I see.

There are several problems regarding honeybee queens, as well as what may be associated problems, but they can largely be put into three groups:

1 Young queens failing.
2 Young queens being superseded.
3 Queens "disappearing".

In addition there are a few other problems that I have seen on a number of occasions. These have appeared fairly recently and may be connected, some of which are mentioned later.

A Bit of History.
In my earlier years of beekeeping there were very few problems with queens. About the turn of the 21st century I started to observe several issues with queens and their performance that I had rarely seen before.

What I write about below seems to be quite universal and is becoming recognised by a growing number of beekeepers, not only in the UK but elsewhere. I gave a lecture on the "Queen Problems" at the SICAMM conference in Switzerland in September 2012. This was an international event with beekeepers of all abilities from all over Europe. Afterwards I had a queue of people wanting to speak to me, to tell me they had experienced exactly the same problems I had described. One French researcher said she was pleased she had heard my lecture, because she had been seeing the same problems for several years. She had discussed it with her colleagues who said she must have been doing something wrong!

At the National Honey Show in 2013 I spoke for some time to a beekeeper from Vermont who had 1500 colonies. He told me he had exactly the same problems I had described.

I have been in discussion with other beekeepers and they tell me they have the same problems. They are widespread, so why is there so much denial? I think there are several reasons. Standards of beekeeping are quite low, so many beekeepers simply don't understand what should be happening in a colony. I'm surprised at the number of beekeepers who have been keeping bees a reasonable length of time who just don't know the "basics". These are the simple things we need to know in order to manage our bees reasonably well.

Many people have come into beekeeping since these problems appeared, so don't recognise them as a problem, in very much the same way as varroa - they think that what they see is the norm. This is becoming more evident when I lecture on the subject, as with the passing of time there are fewer people who agree that things aren't as they once were. I often ask if the audience have seen the problems I have been highlighting. There is usually little reaction until I start explaining what I have seen, then heads start nodding or they whisper to the person next to them. They do see it, but they think it is normal.

I often look back at what happened to bees before our intervention. I have a lecture "Honey Bees in the Wild - What Can we Learn From Them?" and the page on "Natural Honey Bees Nest" where I discuss their survival. Honey bees must have had to keep a fairly stable population based on the process of natural selection. They could not have survived for long if they suffered the queen problems we see today, many of which are terminal for a colony.

What Should Happen?
A queen, depending on her prolificacy, should live for perhaps 3-5 years. I have had many that have managed the latter or beyond. I have rarely culled a queen due to age, as the type of non-prolific queens I prefer will normally perform well throughout their lives. Depending on a number of circumstances, a queen will swarm, or attempt to, perhaps 1 - 3 times during her life. Some won't at all, yet others, such as Carniolans may swarm twice in the same year. She should be superseded in late summer or autumn and very often still be in the colony alongside her daughter the following spring. This is what happens naturally and is in all the old books.

What is Happening Now?
Young queens failing. This is very often in their first year, with many showing signs of failure within weeks of starting to lay, although initially they may appear to be performing well.

Very often a young queen will lay drone eggs in worker cells for a short time before settling down, but I have seen many colonies where there are initially a few drones in worker cells, perhaps 4 - 6 per side of comb. This raises concern with me as the numbers usually increase. I have seen as much as 25% of drones in worker cells. In one case the beekeeper, who had been keeping bees for several years, didn't recognise there was a problem. When I pointed out the single drones in worker cells he said "they are always like that"!

Young queens that are performing well should lay the correct number of eggs compared to the size of the colony, the time of year and the amount of food available.

If you see a full colony in the summer when there is a nectar flow and the queen is only laying on perhaps 2 - 4 frames, with the brood scattered, or the pattern of eggs is poor, when at the last inspection all appeared to be well with brood across the box, then you know the queen is failing.

In my experience once you first see the above the queen usually only lasts about 6 weeks before being superseded, or she "disappears".

Young queens being superseded.
This is often in their first year and I have regularly seen supersedure cells started before the young queen's first brood is sealed. Very often the brood appears to be good to my eye. The cutting out of these cells usually results in others being built and, if continued, my experience is the queen will usually fail or "disappear" in about 6 weeks.

Supersedure cells can be built anywhere on the frame or comb, not always on the face as we are often told. I find they are very often on the periphery of the brood nest and on several occasions I have seen them on combs with no brood on. Individually supersedure cells look like swarm cells, but you can determine them by the quantity. I have a saying of "usually one, often two and occasionally three". Any more than that and they are likely to be swarm cells.

Colonies will swarm on supersedure cells and this is a major problem, especially if there is only one that is on a comb towards the outside of the brood box that has been missed. On many occasions I have had a call from a beekeeper saying they have had a colony swarm that hasn't got any queen cells. I advise inspecting the colony fully, lightly shaking the bees off frames, which often reveals one or more sealed supersedure cells.

When young queens are introduced into a full colony the bees will often build supersedure cells, especially if they are introduced to a colony where the previous queen was laying well. This is common where they were mated in mini-nucs or have been banked for some time. In these cases I assume the bees realise the queen isn't up to speed, so try to replace her. The removal of these supersedure cells usually allows the colony to settle down. This is not what I'm indicating, as this is usually overcome when the queen comes fully into lay.

Queens "disappearing".

This is odd, as there seems to be no sensible explanation. I clip my queens and normally do 14 day inspections. At the Wisborough Green BKA teaching apiary we meet every 10 or 11 days. Good records are kept. You can inspect a colony and find no problems, with no supersedure cells and no problem with the quantity or visual quality of the brood. At the next inspection you can find the queen will have stopped laying instantly, but in about 50% of cases there will be emergency cells, the other 50%, nothing. A check on the age of the brood very rarely suggests the queen may have been damaged at the previous inspection.

The lack of emergency cells may be significant as it suggests the queen has gone off lay, but stayed in the colony for several days, so there are no larvae young enough to be converted into emergency cells. This is one reason I think there might be some "interference" with pheromones as there are other unexplained things happening that are possibly pheromone related.

Other Issues.

At one time a prime swarm could be collected, hived and left for the rest of the season to build up with little attention as it would naturally. Prime swarms would usually have a fertile queen that would last at least until the end of the season, when perhaps she might be superseded. Currently there are many more prime swarms with virgin queens in than there used to be. Very often those swarms with fertile queens will have the queen superseded or she will fail soon after hiving. I believe the former may be caused by the colony swarming on an emergency cell, where perhaps the queen has "disappeared", the latter where a colony has swarmed on a supersedure cell and the queen that has gone with the swarm has soon failed.

There are many reports of "queenless swarms". I suspect there is a queen there, but she has already stopped laying before swarming and doesn't come into lay again.

Queen introduction is nowhere near as simple as it once was. At one time you could take a queen out of a colony, put another queen in a cardboard matchbox or a queen cage with 3-4 workers and place her in the colony immediately. If you went back 24 hours later you could release her, or she would have been released by the bees and there would be no further problems. It was almost unheard of to see emergency or supersedure cells a few days later as often is the case now.

On several occasions I have seen both swarm and emergency cells in a colony at the same time. Unless there has been a manipulation by the beekeeper, such as the queen and some swarm cells removed, this shouldn't happen.

In the past a test comb was very reliable, but in the last few years I have come across several instances where the colony is clearly queenless and they won't build emergency cells (or accept queen cells).

At an apiary meeting of the Wisborough Green BKA on Wednesday 25th June 2014 we inspected 28 colonies and 10 of them had at least one of the problems I mention above. Several colonies had queens laying, yet there were emergency cells in the colonies. The following day I inspected another and found that it had a problem too. This is not a small sample and within shouting distance of 40% of colonies with problems. This is not normal.

There is absolutely no doubt in my mind there are problems and at a guess there may be several causes. The really sad part is that despite there being widespread problems nobody seems to want to do anything about it. In my opinion there is an excellent opportunity for a research establishment to do some meaningful research that may help all beekeepers. I probably know more than anyone on the subject and I would be more than willing to help.

I don't know how many times I have heard that queens aren't getting mated properly because of bad weather. In my experience the same problems arise when the weather is good, but these people don't have an answer for that. I don't think we understand enough about mating in general to make sweeping statements like this. I understand that all bees are capable of forming drone assemblies, but only the native bee is capable of what is called Apiary Vicinity Mating (AVM), so perhaps the answer might be to use native bees.

The success rate of getting a queen mated has reduced considerably. At one time, once you saw a queen cell that had been vacated you could leave a colony for 2-3 weeks and when you went back there was usually good brood. The usual answer if a colony becomes queenless after emergence is the "birds" took them on their mating flights. I'm not buying that one. I think it may happen in a very small number of cases, but not at the rate it is now. Just think of the number of bees there are likely to be on the wing when a queen goes out to mate. The mathematical chances of a queen being taken are very low. I strongly suggest everyone looks at the wings of virgin queens as there are now a significant number that emerge with deformed wings. This varies from tiny stubs to almost full wings that are crumpled at the tips. I think they may emerge from the hive to mate, but don't get airborne. This suggests a possible virus problem, so perhaps we should look in that area.

What Are the Reasons?
I am an engineer by trade, not a scientist, so I will try to leave the answers to others. Amongst the more sensible suggestions that have been made to me, though not in any order are:-

- **Chemicals.** That are administered both inside and outside the hive.
- **Disease.** Varroa and associated viruses and queens infected with **nosema.** Unless the nosema is *Nosema ceranae*, I think this is doubtful, as we have always had *Nosema apis* and we never used to have the problems.
- **Bad beekeeping.** Good beekeepers have the problems and there have always been bad beekeepers.
- **The weather.** This always gets the blame, but queens are still failing when they get mated in good weather.
- **Mobile telephone masts.** I think these may be too quickly dismissed by some.
- **Inbreeding.** There are a number of locations where there are small closed populations. Some are regularly monitored and there are few problems. Bearing in mind the density of colonies in most locations and we are told there are drones from 200 - 250 colonies in many Drone Assemblies, I think inbreeding is extremely unlikely.
- **Poor drone quality.** We know that drones that have been parasitised by varroa and subject to some chemicals have a lower sperm count. Drone culling and the reduction of feral colonies is blamed by some, but I think there are still enough drones.

Some of My Guesswork.
I am an observant beekeeper and with my engineers mentality I use a bit of logic. What follows is something I have been thinking about for some time. I must stress it is no more than a non-scientific theory, but please feel free to pick as many holes in it as you like.

Many of the problems could be pheromone related. I have done no reading on them, but with my practical engineers mind I see a pheromone as a substance that is made up of a number of components, each one having a different percentage of the whole. I'm guessing that if one or more components has a different amount than it should have, or is missing, it gives a different message to the worker bees.

Pheromones are chemicals and chemicals can be altered, something we all know from our school science lessons. My thinking is there may be some sort of "interference" with the pheromones, perhaps with chemicals that are coming in from outside the hive, or those administered to the colony. If that modifies the pheromone it may give the wrong message to the worker bees. I hope I explain myself adequately.

If the above is close to being a possibility it may explain some of the things that are happening with queens.

- The queen is normally fed by the workers, presumably because she is producing a pheromone, so they recognise her as a queen. If the pheromone changes, so they don't recognise her as a queen, could the bees not feed her, so she starves, hence what I see as "disappearing"?.
- If the pheromone changes slowly the queen presumably gets fed progressively less food, so her egg laying reduces and this is why she appears to be failing.
- If the pheromone reduces could it be why the colony build supersedure cells? When queens are fed they receive queen substance. If that is at a low level it triggers off the laying of eggs in queen cups.
- Where I have noticed less success with queen introduction, or where recently introduced queens are quickly superseded, are the pheromones already reducing and the queen would have been superseded if left in the original colony?
- I'm baffled by swarm and emergency cells in the same colony, because there is a different message being given to the workers. I haven't thought too much about that yet.

I accept the above is conjecture, but in the absence of any other explanation lets chuck it in the pot with everything else. One thing is for sure and that is it's a better bet than keep on blaming the weather and birds.

Why Has there Been Little or No Progress?

I have tried desperately hard to make beekeepers aware of the above problems over a long time and I have become very frustrated at the lack of progress. I have tried to highlight them, but I have experienced negative reactions from people in influential positions who I think should know better. Quite frankly some simply don't believe me and dismiss what I tell them that I have seen with my own eyes. I have come to suspect that some well known beekeepers may not be as knowledgeable or observant as their status in beekeeping suggests. I wonder how often they inspect their own bees or if they understand what they are seeing.

On several occasions I have been told by beekeepers they didn't have a problem, but when I ask to look at their bees, I often see some. As always when some people don't know much about a subject they rubbish it, ignore it or try to discredit it. I have been asked to provide proof there were previously no problems, but who records things when they go well?

The negativity or lack of knowledge of some beekeepers probably has a bearing on it. I have been openly told that the reason I have problems is because I'm a bad beekeeper, which I find unhelpful and offensive. I am doing the same as I have for 50 years and I'm more than happy to inspect colonies in front of anyone.

What Can be Done?

I think beekeepers can do quite a bit on their own, but it needs to be in an organised way, not a scattergun approach. I am not a biologist and know little about the anatomy of bees, but I think beekeepers who have an interest in microscopy can dissect queens to study spermatheca and ovaries. If there is variation from the normal healthy organs this could provide data for researchers to take further.

I am out of my depth already, but my hope is that any abnormalities could be tested for diseases, viruses, poor sperm, etc.

My thinking is that queens in fairly high numbers are easy to produce during the summer, so could be used in a number of ways, as virgin and fertile queens. I am prepared to do this myself and I think I know a number of queen rearers throughout the country who would also be prepared to provide material.

There are a number of locations that are still varroa-free, so queens reared there could also be dissected to see if they have problems.

For the sake of beekeeping I hope someone can at least do some exploratory work. I'm happy to provide any information I can. I also give lectures on the issues, so if you want to know more please contact your local BKA secretary.

Roger Patterson is the author of 'Beekeeping - A Practical Guide' - available in print or digital format.

WINDSOR HILL WOOD:

A beekeeper's sanctuary provides a refuge for people in need of care.

Tobias Jones

Six years ago, my wife and I set up a residential woodland community for people in a period of crisis in their lives. The 10-acre site was an old quarry in Somerset which had been abandoned 50 years before, and over the subsequent decades it had become an enchanting woodland, albeit a post-industrial one: there were bits of clinker and metal and builders' rubble amongst all the oaks, ashes, hazels and hawthorns. It was so far from any major road or city that we frequently saw deer and rabbits walking right up to our house, which was the old quarry-master's pad.

Windsor Hill Wood quickly became a renowned refuge for a whole range of people – for those suffering, typically, from addiction, bereavement, depression, PTSD, eating disorders, homelessness and so on. Having started with just one guest at a time, we soon had half a dozen people living with us, most staying for six months or more in various shepherd's huts, yurts, camper vans and outbuildings. The fact that we were a strictly dry and drug-free house, with a lot of manual labour, good food, discipline and love, somehow seemed to help people get their lives back on track.

We started breeding pigs and sheep and chickens; growing industrial quantities of vegetables, heating our house and its water with our own wood, making charcoal and creating bespoke chairs and tables in our workshop. But we made plenty of mistakes - it was often the blind leading the blind: we put the poly tunnel in the wrong place (with poor soil and too much shade) and had to take it apart and start from scratch; we were very bad at fencing to begin with, and the pigs would often escape and plough through our precious vegetables; and I wasn't, in the beginning, a very confident leader, and would give disruptive people a second chance, and then a third, and a fourth, when I should have put my foot down, asserted the rules and told them to leave.

But by far my greatest area of incompetence was with our bees. All my life I had watched my father among his hives and pottering about in his bee shed. He's a tall man and in that white suit, with a black veil over his face, he looks a bit like an FBI baddie from *ET: The Extra Terrestrial*. The rich aromas of beekeeping always remind me of him: the nurturing smell of wax and the soothing white plumes from the burning cardboard in his smoker (a sort of bellows to calm the bees).

I had seen the way that his beekeeping made him friends wherever he went: it wasn't just that he would give wax or honey to people instead of a dull bottle of plonk, he would also (as he was the South Somerset "swarm man") get called round to peoples' houses when they had a swarm, and – having captured the colony – would normally stand around chatting to them, making connections.

And my old man could make a connection with anyone. One of the pleasures of beekeeping, it seemed from a distance, was that it was able to break down all sorts of barriers of class and age. It brought together everyone, from earthy farmers to aristocratic landowners, from feisty old women to enthusiastic teenagers. Dad is so gregarious, and generous, that he seemed to pull all those people together, not just around bees but around himself too. Even now, it's not infrequent for someone from the far side of the country, applying to come to live on our little sanctuary, to ask me: "You're not related to Bob the beekeeper, are you?"

Bees were what made my father a much-loved local eccentric. For years he used to put a hive on a busy roundabout in Wincanton because it was covered with wild-flowers and dandelions. Most towns have some plaque declaring that a roundabout is sponsored by some local business; in Wincanton they've got batty "Doctor Bob", the town's GP for 35 years, and his hives. He was stopped on that roundabout by the police late one night, as he went through the frames under the A303. The local officer recognised the bald, Norman Tebbit-look-alike and smiled to his colleague: "Don't worry – it's only Doctor Bob."

So, like all men who admire their fathers, I probably wanted to be a bit like my old man and beekeeping was my method of imitation. But there were other